A Great Idea
ENGINEERING

The Large Hadron Collider

by Bonnie Juettner Fernandes

Content Consultant
Dr. Don Lincoln
Senior Scientist at Fermi National Accelerator Laboratory

NORWOOD HOUSE PRESS

COVER: The Atlas experiment is conducted at the CERN Hadron Collider in Geneva, Switzerland.

Norwood House Press
P.O. Box 316598
Chicago, Illinois 60631

For information regarding Norwood House Press, please visit our website at:
www.norwoodhousepress.com or call 866-565-2900.

PHOTO CREDITS: Cover: © Massimo Dallaglio/Alamy; © Anja Niedringhaus/AP/Corbis, 38; Bow Editorial Services, 8, 11, 26, 35; CERN, 12, 27, 31; © CERN/Science Source, 32; Daily Mail/Rex/Alamy, 15; © Denis Balibouse/AFP/Getty Images, 37; Edal/Wikipedia Commons, 19; geni/Wikipedia Commons, 17; NASA, 6, 9; © Richard Wareham Fotografie/Alamy, 23; © Science Source, 42; © SSPL/Getty Images, 21, 30; © Ted Foxx/Alamy, 15; Wikipedia Commons, 41; © World History Archive/Alamy, 13

Paperback ISBN:978-1-60357-580-5

The Library of Congress has cataloged the original hardcover edition with the following call number: 2013012254

Manufactured in the United States of America in North Mankato, Minnesota.
233N—072013

Contents

Note: Words that are **bolded** in the text are defined in the glossary.

Chapter 1

What Is the Large Hadron Collider?

The Large Hadron Collider (LHC) is a tool that scientists use to learn more about how the **universe** began. The universe includes Earth, the solar system, the Milky Way **galaxy**, other galaxies, and the stars that can be seen. It even includes the stars that are not seen but that scientists know exist. The universe is billions of years old. It is older than any fossil. It is older than the dinosaurs. It is older than Earth and older than the sun. But even

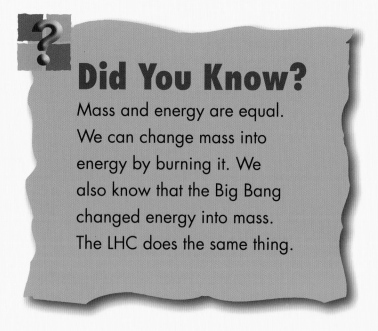

Did You Know?

Mass and energy are equal. We can change mass into energy by burning it. We also know that the Big Bang changed energy into mass. The LHC does the same thing.

Before the Big Bang

What happened before the Big Bang? Scientists can only make educated guesses. Albert Einstein did not believe that time existed before the Big Bang. He thought the Big Bang was the beginning of time. But some scientists do not agree with Einstein. They argue that before the Big Bang, there was another universe that had a Big Crunch. Perhaps the universe contracted before it expanded. Other scientists think that our universe may be a bubble universe. It may be located inside a larger multiverse that contains other bubble universes. Still other scientists have thought that our universe somehow came from a black hole in another universe. Nobody knows for sure.

though our universe is very old, scientists believe that it had to have a beginning.

Scientists think that billions of years ago, all of the many parts that make up the universe were inside a tiny point. This point was smaller than the tiniest speck that you could draw with a sharp pencil. It was smaller than the tip of a hair. The universe was inside a point so small that not even the strongest microscopes today could have detected it. It was so small that it may not have even been **three-dimensional**.

This tiny universe was made of energy. It was very hot. It was much, much hotter than our sun is today. Scientists think that the tiny universe started to expand and cool after a sudden explosion called the **Big Bang**. As the universe expanded, it became cooler.

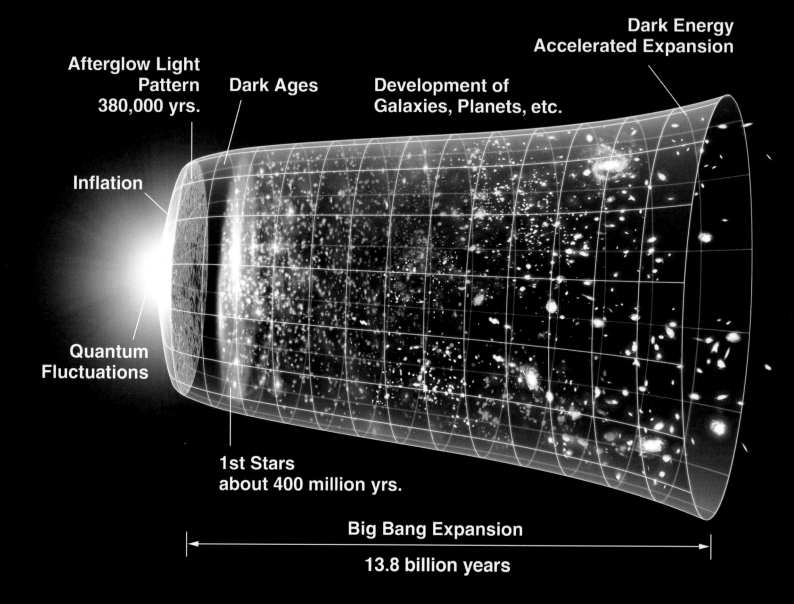

Afterglow Light
Pattern
380,000 yrs.

Dark Ages

Development of
Galaxies, Planets, etc.

Dark Energy
Accelerated Expansion

Inflation

Quantum
Fluctuations

1st Stars
about 400 million yrs.

Big Bang Expansion

13.8 billion years

Once the universe had expanded for 400,000,000 years, stars began to form from big clouds of gas. The stars formed in groups called galaxies.

Matter

Over time, scientists have learned a lot about the Big Bang and the universe. They know how the universe's temperature changed over time. They also know how long it took for stars and planets to form. They also understand how matter is formed. Matter is any substance that has weight and takes up space. But the early universe, just before the Big Bang, did not take up the same space that matter in our universe takes up now. If it

Timeline of the universe. 400 million years after the Big Bang, stars began to form and spread out into groups called galaxies.

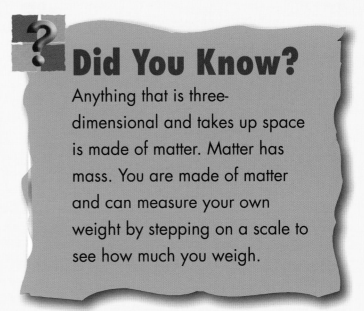

had, it would never have fit into such a tiny point. Where did the first matter come from?

Peter Higgs

In 1964 British **physicist** Peter Higgs went for a walk. He came back thinking about the ocean. When people swim in

The Building Blocks of Matter

All matter, including the cells that make up your body, is made up of atoms. Atoms are too tiny to see with even the best microscope ever made. What do atoms look like? Atoms have a nucleus made of two types of particles. These are called protons and neutrons. Both protons and neutrons are types of hadrons. Other types of particles orbit the nucleus in a cloud. These are called electrons. Protons and neutrons are made up of smaller particles.

In between the particles of the atom is empty space. The atom is almost entirely empty with 99.99999999% of it being empty space! If you were tiny enough—smaller than an atom—solid things would stop looking solid to you. You would be able to see the spaces between an atom's nucleus and its electron cloud.

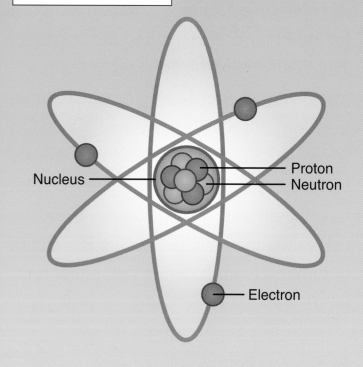

Atom Diagram

Nucleus

Proton
Neutron

Electron

British physicist Peter Higgs first thought of the idea of an ocean in space—the Higgs field—that contained a special type of particle, the Higgs boson, in 1964.

an ocean, they get wet. Higgs imagined a kind of ocean in space— but not a water ocean. He imagined an ocean that, instead of making particles wet, could give them mass. Today the ocean that Higgs imagined is known as the **Higgs field**.

But Higgs wondered why different particles of matter have different mass. So he also imagined that his ocean, or Higgs field, must contain a special type of particle. This is now called a **Higgs boson**. Higgs thought that if particles interacted with the field he imagined, they would

Higgs's Party

Today scientists have a way to explain the idea of the Higgs field and Higgs boson. They tell people to imagine a party of students who are uniformly distributed across the floor, all talking to their nearest neighbors. A former teacher enters and crosses the room. All of the students in her class are strongly attracted to her and cluster around her. As she moves, she attracts the students she comes close to, while the ones she has left return to their even spacing. Because of the knot of students always clustered around her, she acquires a greater mass than normal. If an unpopular teacher crossed the same room, the students wouldn't cluster around him and he could move freely. He would be like a particle with no mass.

gain mass. Particles that interacted a lot with the field would be very massive, while those that didn't interact with the field wouldn't have any mass at all. The Higgs field has big impact on the mass a particle has.

Over time, Higgs's ideas were accepted by physicists all over the world. Still, to be sure that the Higgs boson even existed, scientists needed proof. They needed to find and observe the Higgs field or a Higgs boson.

Re-creating the Big Bang

In the mid-1980s scientists thought they had found a way to look for the Higgs boson. They thought they needed to use particles as small as the parts of an atom. These are **hadrons** (which can be protons or neutrons) and **electrons**. Scientists

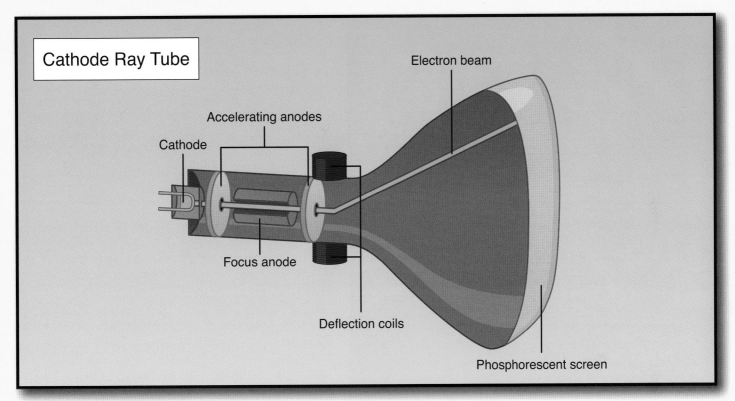

Cathode Ray Tube

Electron beam

Accelerating anodes

Cathode

Focus anode

Deflection coils

Phosphorescent screen

In a cathode ray tube a beam of electrons is accelerated to 30,000 volts. As the electrons pass across the television screen, or monitor, they cause the screen to give off light.

believed that they would need a **particle accelerator** to do the job. The earliest particle accelerator was a **cathode ray tube** (CRT). The CRT was used in TVs and computers until the 21st century. In a CRT, a beam of electrons is sped up to 30,000 **volts**. As the electrons pass across the TV or computer screen, they cause

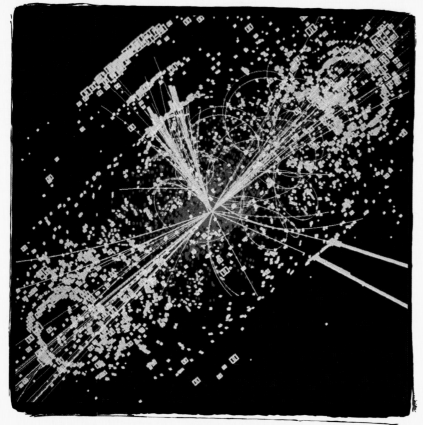

A Higgs boson from a simulated collision between two hadrons. It decays almost immediately into two jets of hadrons and two electrons, visible as lines.

day and can accelerate both electrons and hadrons. Only about 110 of them are used for particle physics research.

Scientists believe it is impossible for a particle to move at the speed of light. But physicists wondered what would happen if they could speed up particles such as protons and electrons to very close to the speed of light. Neutrons are hard to accelerate because they have no electric charge. Physicists wondered what would happen if they not only sped up particles to almost light speed but then crashed, or collided, them together. They hoped that the heat and energy

the screen to give off light. Today particle accelerators are used all around the world to make all sorts of images. About 15,000 particle accelerators exist in the world to-of two particles crashing together might be strong enough to reproduce, on a very small scale, the heat and energy caused by the Big Bang. If so, scientists could watch what happened. They might even discover new particles, like the Higgs boson, in the process. The experimental Big Bang would not have to be as large as the one that created the universe. It would only have to be large enough for scientists to observe what happens—and to search for signs of the Higgs boson.

This is a particle accelerator. Particle accelerators speed up particles to almost the speed of light and crash them together to reproduce (on a small scale) the heat and energy caused by the Big Bang.

Chapter 2

Challenges and Solutions in Developing the LHC

Scientists first began thinking of building the LHC in the 1980s. By then physicists had been working on particle accelerators for almost one hundred years. In fact, scientists had been working with accelerated particles since the discovery of **electricity**. Scientists knew how to move particles called electrons through wires. This is kind of like how a particle accelerator works, but a particle accelerator does something even cooler. Instead of moving the particles

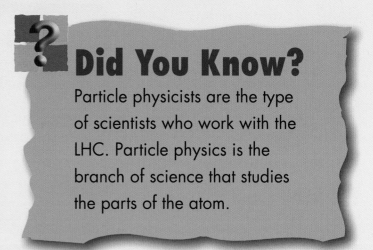

Did You Know?

Particle physicists are the type of scientists who work with the LHC. Particle physics is the branch of science that studies the parts of the atom.

The ATLAS is a particle detection device and experiment at CERN that detects and studies particles and their movement.

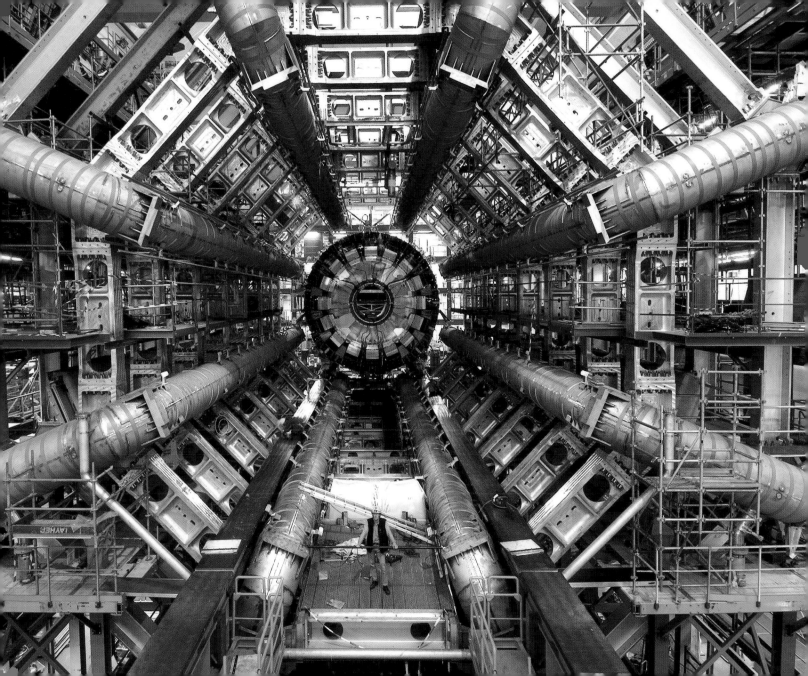

The Vacuum Seal

If air gets into a particle accelerator, the air molecules could bump into the protons or electrons as they speed up. This would knock these particles off their paths. So a particle accelerator must be vacuum sealed. All the air has to be taken out of the tube or chamber.

through a wire, particle accelerators do something trickier, they can cause particles like electrons or protons to travel through empty space.

Your Living Room Particle Accelerator

The CRTs used in older TVs also used an electric field to speed up particles. The CRT is a sealed tube. The tube also in-cludes a filament like those in lightbulbs. When a TV with a CRT is plugged in and turned on, electricity heats the electrons in the filament which jump out of the fila-ment and are then accelerated by electric fields. They stream from one end of the tube to the other at high speed. One end of the tube is coated in phosphor. This glows when electrons hit it. An old TV also con-tains magnetic coils. These steer the beam of moving electrons so they "paint" a pic-ture on the screen as they move.

As strange as it may seem, a CRT is like a mini version of the LHC. The LHC is much bigger. It uses much more energy. It also requires much larger—and colder—magnets. Also, the LHC does not speed up electrons. Instead, it speeds up proton particles from the atom's nucleus.

This early Cockcroft-Watson accelerator could accelerate particles to very high velocities.

Accelerating in a Straight Line

Scientists still needed something before the LHC could be built. They needed an accelerator that could make much more energy than a CRT. In 1932 John Cockcroft and Ernest Walton had an idea. They invented the Cockcroft-Walton accelerator. This makes about 30 times as much energy as the electron beam in a CRT. It works by pushing the electrical current through a series of steps. At each step, the **voltage** increases. This type of accelerator is very powerful. But the

Cockcroft-Walton accelerator and other similar accelerators had a problem. They moved particles in a straight line. The particles moved down a long tube. They moved faster the farther they went in the tube. So, to get particles to go very fast, the tube had to be very long. Eventually, the tubes could not be made long enough to be practical for scientists.

The First Cyclotron

Ernest Lawrence called his first particle accelerator his "proton merry-go-round." Later, other scientists dubbed it the cyclotron. The first cyclotron was shaped like a pie. It was made out of glass, sealing wax, and bronze. Lawrence also used a kitchen chair and a clothes tree made out of wire coils as part of his design.

The Cyclotron

About the same time that Walton and Cockcroft were working on their accelerator, Ernest Lawrence was working on the same problem. But instead of forcing

A photo of Ernest Lawrence with an early version of the cyclotron he invented.

Paying for the Cyclotron

Physics experiments can be costly. Lawrence first began working on the cyclotron during the Great Depression. To save money, he used an 80-ton (72.6t) leftover magnet that had been used to power a radio link across the Atlantic during World War I. He used volunteer physicists who could not find jobs anywhere else. And he did medical research with the cyclotron to help pay the bills.

his particles to speed up in a straight line, he had another idea. He decided to try moving the particles along a spiral path.

Using a spiral meant that the particle accelerator could take up much less space. The particles could pass the same point many times. This is because they would move along a curve. In Lawrence's model every time particles passed a certain point, they received more electric voltage. This made them move even faster. This type of accelerator takes up less room. It gives the beam more energy. To keep the particles moving along the spiral path, Lawrence used magnets—just as the LHC does today. In 1939 Lawrence won the Nobel Prize in Physics for his invention. He called it the cyclotron. Accelerators that work in a similar way today are often called synchrotrons. Synchrotrons make particles move not in a spiral, but in a circular path.

The Bubble Chamber

The cyclotron worked very well as a way to accelerate particles to very

Physicist Donald Glaser works on his bubble chamber at the University of Michigan in 1952.

high speeds. But scientists were still having trouble doing experiments on particles and observing the results. They needed a structure that could hold the particles while pictures were taken and data was gathered. Some say that Donald Glaser came up with an idea while he was drinking a glass of beer. He watched the bubbles in the beer rise to the top of his glass. He thought it might be possible for the particles to cross the liquid and leave a path of bubbles. You could then photograph the bubbles and know where the particles went.

Glaser invented a bubble chamber in 1952. He, too, won a Nobel Prize for his work. Over time, scientists tried many kinds of liquids in their bubble chambers, including hydrogen gas cooled so much that it became liquid.

The bubble chamber is a tank of liquid. As the sped-up particles arrive in the chamber, their electric charge causes the liquid to boil. The boiling liquid forms bubbles. Scientists let the bubbles grow, and then they take pictures. Then they clear the bubbles from the chamber and wait for another stream of particles. Sometimes the experiment is repeated hundreds of thousands of times. Over time other technologies have been developed that have replaced the bubble chamber.

Superconducting Magnets

Scientists then started working more with accelerators. They soon found that

A superconducting magnet used at the LHC. It accelerates protons, particles from an atom's nucleus.

they needed a way to keep the sped-up particles moving. It was not enough just to use electric fields to accelerate particles to high speeds. And the particles had to be kept in their paths. If they were not, they might start bouncing around the accelerator instead. Lawrence was one of the first scientists to use magnets to keep particles in their paths.

Today the LHC has about 9,600 magnets. Some of these magnets keep the particles on track in their paths. There are always two streams of particles moving in opposite directions—that way, the magnets can shift the beams until they are aimed straight at each other. Other magnets work to squeeze the beams together just before two particle beams are aimed at each other. Squeezing the particle beams together brings the particles closer and closer to each other. Eventually, they are so close together that the particles in one beam are forced to collide head on with the particles coming from the other stream. To do this, the LHC does not use ordinary magnets. Ordinary magnets, even when they are as big as a truck, are not strong enough to hold sped-up particles in their paths. Instead, the LHC uses very cold magnets called **superconductors**. These magnets have to be cooled with liquid nitrogen and liquid helium until they reach a temperature that is colder than outer space: -456.3°F (-271.3°C).

Chapter 3

Constructing the LHC

Scientists had solved the technical problems in the design of the LHC. But they still had to build it. Digging the tunnels and caverns that were needed to house the LHC was a huge job! The space had previously held an earlier version, the Large Electron–Positron Collider. The LHC was the largest project Europe's engineers had taken on since the Channel Tunnel. This is the tunnel under the English Channel.

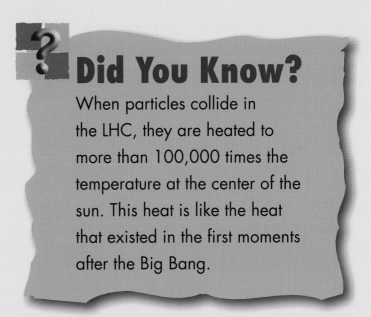

Did You Know?

When particles collide in the LHC, they are heated to more than 100,000 times the temperature at the center of the sun. This heat is like the heat that existed in the first moments after the Big Bang.

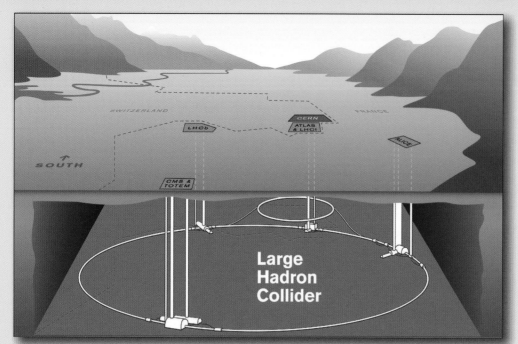

The Large Hadron Collider is housed in a 16.8-mile-long circular tunnel near Geneva, Switzerland.

the LHC was already in place, much **excavation** work was done to add additional experiment sites starting in 1998.

Many nations were involved in building the LHC. It is underground. It is in a tunnel that is 16.8 miles (27km) long. The tunnel is beneath the border of France and Switzerland, near Geneva. The CERN laboratory, the European Council for Nuclear Research, has been located there since 1954. Although the tunnel for

Step 1: Excavation and Construction

Workers began to dig the soil at the LHC site to add the CMS detector, one of the experiments. But they soon found a surprise. There was an ancient Roman villa beneath the soil. Work on that end of the site ceased for a year while **archaeologists** explored the find.

Then, in the same area, workers faced another challenge. They found water 33 to 66 feet (10m to 20m) below the surface. They had to dig through that water. First they froze it into a 10-foot (3m) wall of ice. But water was still coming in faster than they could freeze it. Finally they injected the area with liquid nitrogen. This

The CERN complex outside Geneva sits atop part of the LHC.

International Cooperation

The LHC is an international project. Many nations have worked together to help make it possible. They also help pay for it. The list of nations that have helped build the LHC includes the United Kingdom, Germany, France, Switzerland, and sixteen other European countries. It also includes the United States and Japan. It cost about 8–10 billion dollars to build the LHC.

is colder than -319°F (-195°C). The ice formed walls of ice around the shafts that engineers were digging, protecting the shafts from being flooded. Between 1998 and 2005, workers at the LHC site removed almost 8.8 million cubic feet (250,000 cubic meters) of soil and rock.

The LHC Caverns

At this point, workers were ready to build the vast caverns that would hold the LHC and the four new experiments: CMS, ATLAS, ALICE and LHCb. But now they faced a new problem. The ground in the CMS area was too soft. It was made of glacial deposits called moraines. These are formed from a mixture of sand and gravel. A cavern dug in this material would collapse. Engineers solved this problem by building a concrete supporting structure. It would hold the caverns up and support the weight of the soil above them. But they had to move slowly, one small section at a time. As a small part of the cavern was dug, they would quickly spray it with "shotcrete." This is a fast-drying type of concrete. It sets almost immediately once it is sprayed on the walls. Then they drilled steel anchors for the section, 39 feet (12m) into the surrounding rock. If they had not done so, the engineers feared that the whole cavern would collapse while they were building it.

While they were building, engineers tried to consider the people around them.

They put up sound barriers to protect people from the noise of the digging. And they did not move the huge piles of rock and dirt they dug out. Instead, they just put the material to one side and replanted it. This made an artificial hill.

While some engineers were digging the site, others were above the surface. They tested the magnets as they arrived. They also assembled the many pieces of the LHC. The LHC and the experiments consist of many, many parts. In 2005 engineers finished digging and building the tunnels and caverns.

Step 2: Collecting the Data

When the construction was complete, scientists now had new ways to collect data from the LHC. They made four sets

CERN scientists celebrate the first collisions in 2010.

of tools for measuring the conditions that existed just after the Big Bang. These are called ATLAS, ALICE, CMS, and LHCb.

ATLAS is a particle detection device and the experiment that it is used for needs its own team of scientists. ATLAS can detect particles and study their movement. It also has a calorimeter. This is a tool that absorbs particles and then measures each particle's energy. And it has a tool called a muon spectrometer. This is a

The ALICE experiment has a special tool called a Time Projection Chamber (shown) that studies the path that a particle travels inside the LHC.

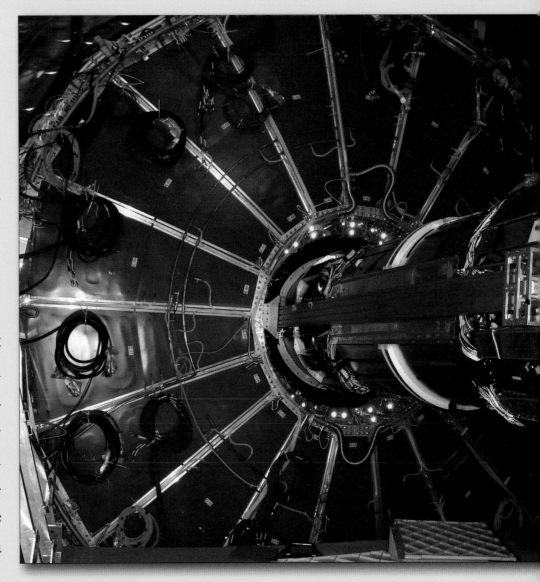

device that can detect a special particle called a muon. Muons are easy to see and live longer than most subatomic particles.

Studying Collisions

The ALICE experiment studies collisions between ions, atoms that have a positive or negative electrical charge. It has a special tool called a Time Projection Chamber. This can be used to study the path that a

The Mystery of Quarks and Gluons

Scientists have named the tiny particles that protons break into when they collide. They are made of smaller particles called quarks. They are held together by gluons. Gluons transmit a force that is strong enough to hold two or more protons together. They can do this even though protons all have a positive charge. Electrically, two positive charges should repel each other. But in the nucleus of the atom, gluons are able to hold these like charges together. How they can do this is a curiosity. Another curiosity is that if all the quarks in a proton are put together, they still do not weigh as much as the proton does.

A computer-constructed subatomic particle collision. Each yellow line is a particle coming out of the collision.

The LHC and the Moon

Scientists at the LHC have to include the movements of the moon in their equations. The moon's gravity pulls Earth's oceans, creating tides. But it also pulls Earth's crust, causing "ground tides" and changing the shape of the LHC tunnel slightly. Scientists at the LHC adjust their equations during new moons and full moons.

particle travels inside the LHC. ALICE also studies a rare state of matter. It is thought to have existed only in the first moments after the Big Bang. It is called quark-gluon plasma. Ordinary plasma is gas that is so hot that some of its atoms have released, or lost, their electrons. The electrons can then move about freely. You have seen this kind of plasma if you have ever watched a lightning bolt. Plasma is also found inside fluorescent light bulbs. However the type of plasma studied by ALICE is far hotter than that. In these collisions, the temperatures are so hot that not only would atoms be torn apart, but the protons and neutrons at the center of atoms would actually melt and release their **constituent** particles called quarks and gluons. The last time that this kind of matter was common in the universe was about a millionth of a second after the Big Bang.

The Compact Muon Solenoid (CMS) experiment detects and measures particles. When two protons collide, this can cause the protons to break apart

The LHC's Cooling System

The magnets that move the hadrons in the LHC are stored in one of the coldest places on Earth. To keep it cold, first the LHC uses liquid nitrogen to cool down to -451.7°F (-268.7°C). Then cold helium is injected into the magnets. Together they bring it down to -456.3°F (-271.3°C).

into even smaller particles. The CMS team looks for these particles and studies them carefully. To do this, they use a giant magnet. It creates a magnetic field about 100,000 times stronger than Earth's own magnetic field. The CMS experiment is so big that it had to be put together in pieces. Fifteen pieces of CMS were put together above the ground. Then they were lowered into the caverns used by the LHC. Once the pieces were underground, CMS was put back together.

The LHC also has many other experiments. These study other aspects of particle collisions. One of them, called LHCb, tries to figure out why the universe is made of matter. The experiments and measurement tools at the LHC collect a large amount of data. All together, these tools have hundreds of millions of sensors. These transmit data 40 million times per second. About 25 million gigabytes of data come out of these experiments per year. If all this data were put on DVDs, the stack of DVDs made each year would be 12 miles (20km) tall. The

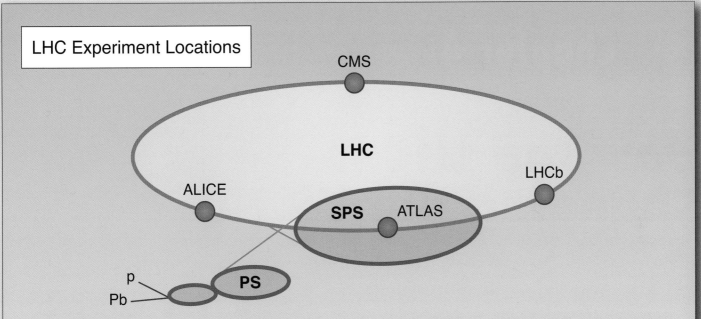

LHC Experiment Locations

CMS

LHC

LHCb

ALICE

SPS ATLAS

p

Pb

PS

For the LHC experiments the path of protons begins at the Linear Accelerators (P and Pb respectively), they continue through the booster (small unmarked circle) to the Proton Synchrotron (PS), the Super Proton Synchrotron (SPS) and through the 16.8 mile (27 km) tunnel where the ALICE, ATLAS, and CMS experiments are conducted.

data actually gets stored on a tape robot. With so much data, computer scientists came up with a new system to analyze the data. This is called the LHC Computing Grid project. It is a network of computers from all over the world.

Step 3: Setting Up the LHC

The LHC was finally installed. It started up on September 10, 2008. Scientists were ready to start figuring out exactly how matter came to exist. They were ready to re-create the Big Bang.

Chapter 4

How the LHC Is Changing the World

For researchers at the LHC, July 4, 2012, was the day that everything changed. Peter Higgs, who was traveling in Europe at the time, got a phone call. John Ellis, the head of theoretical physics at CERN, left a message for him. "Tell Peter that if he doesn't come to CERN on Wednesday, he will very probably regret it," the message said. Higgs went to the meeting at CERN. He got there in time and physicists Fabiola Gianotti who worked on ATLAS,

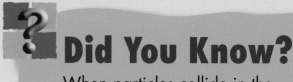

Did You Know?

When particles collide in the LHC, mass can turn into energy and energy can turn into mass.

and Joe Incandela who worked on CMS, began to present their most recent results of the LHC research. They were projecting their notes onto a screen for the audience to see. But the physicists in

On July 4, 2012, physicists Fabiola Gianotti, left, and Joe Incandela announce to a meeting of the scientific community that CERN had uncovered evidence that the Higgs boson may be real.

"It's a triumphant day for fundamental physics," said Princeton physicist Nima Arkani-Hamed. "Now some fun begins."

Recognizing the Higgs Boson

the audience read faster than they could talk. They scanned ahead of the notes and saw a symbol that meant the Higgs boson. The LHC had found evidence that the Higgs boson—thought to be responsible for giving particles mass—probably is real. When they saw this, scientists in the audience began wildly cheering.

In order to show that the particle they had found was a Higgs boson, scientists had to take many precise measurements, to be sure that the particle they had found acted like they expected. The ATLAS and CMS detectors in the LHC are the biggest, most precise measuring tools ever built. They

The ATLAS experiment can detect particles, analyze their momentum, and measure each particle's energy.

can measure the time it takes a particle to pass through space with precision of a few billionths of a second. They can also measure the location of the particles at a precision of a few millionths of a meter.

To start taking these measurements, scientists first set up the LHC and crashed beams of protons together. Once the protons collided, the energy of the collisions made new and sometimes short lived matter that could be observed in the detectors. The ATLAS and CMS detectors kept track of signals left as the collision process unfolded. Just as wildlife biologists can identify animals from tracks left in the snow or mud, physicists at the LHC can identify particles by the signals and traces they leave behind in detectors. It is important for scientists to have data showing signs that a particle was there. This is because the particles produced in collisions do not survive very long. They quickly decay and break apart.

Later, using the data from ATLAS and CMS, scientists were able to reconstruct what happened. Some scientists have said that the process is much like police officers reconstructing a vehicle collision by using evidence from the scene. But unlike a police investigation, the LHC collects so much data that it can take many months to study it all. So even after taking such precise measurements, scientists were still not 100 percent sure

Did You Know?

The LHC could be thought of as the fastest racetrack on Earth. At their fastest speed, two beams of protons race around the rings of the accelerator in opposite directions 11,245 times per second, or nearly the speed of light. About 600 million collisions take place every second.

that they had found the Higgs boson—because they had not yet studied all the data. But CERN was ready to announce in July 2012 that the particle they had found was "Higgs-like." Scientists felt this was okay, because the particle they found had about the same properties that they expected the Higgs boson to have.

Is the LHC Safe?

Some people were afraid that re-creating the Big Bang at the LHC could somehow destroy the world. One of the most common fears was that the LHC could produce a black hole — an area of space that has such strong gravity that nothing can escape it — right here on Earth. But physicists say that the LHC is completely safe. They say that if it did produce a black hole, it would be so tiny that it would disappear right away. The Big Bang made at the LHC is not the same size as the original Big Bang that made our universe. This Big Bang is smaller. It is even smaller than an atom.

Over the next six months, scientists studied even more of the data. They checked each particle's spin. They also checked to see whether the particles moved in a symmetrical way or not. They analyzed the path that all the particles traveled along after each collision. And they noted the ways that the particle interacted with other particles. They studied the data for seven more months. They finally announced in March 2013 that the data "strongly" supported the idea that the particle they found was the Higgs boson.

Practical Uses for the LHC Research

The research that went into making accelerators is already being used in the rest of the world. It is used mainly by doctors in hospitals. The study of particle acceleration helped scientists to develop a type of medical imaging called positron

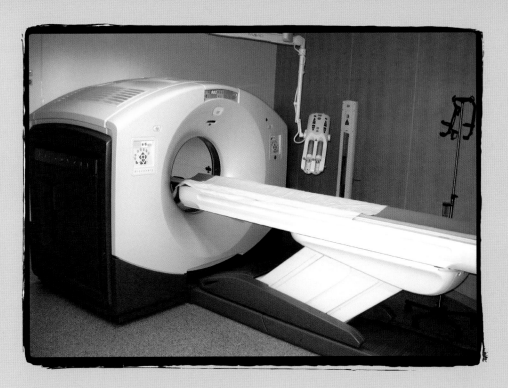

The study of particle acceleration helped scientists develop medical imaging machines called PET (positron emission tomography) scanners.

emission tomography. In addition, doctors are starting to use streams of proton beams like those in the LHC to treat cancer. Cambridge physics professor Andy Parker says, "What you can do is to send a beam of protons into the patient, which does essentially no damage at all to the tissues on the way in … all the damage is done at the point where the protons stop. And by tuning the energy of the protons, you can make them stop inside the tumor." This means that doctors can operate on cancer patients without making a single cut in the patient's body.

Doctors are happy that the LHC research has been useful in so many ways. But other scientists say that even if the research had no practical use, it would still be worthwhile. They insist that the

Dark Matter and Dark Energy

Only 4 percent of the universe is made up of ordinary matter. The rest is thought to be dark matter and dark energy that we cannot see. Scientists hope the LHC will be able to detect signs of dark matter. Dark energy probably won't be observed at the LHC, although scientists will certainly look for it.

An image from the Hubble Space Telescope shows the huge ring of dark matter that scientists are investigating.

most important part of the LHC is that it reveals more about the universe.

What Happens Next?

The LHC has been shut down for a short time. Although it operated for only four years, the technology in the LHC had already become dated and the equipment required maintenence. Scientists also needed to update the LHC with more-modern equipment. The LHC will start up again in 2015. When it does, scientists hope to find out more about what happened during the Big Bang. They want to know what happened to **antimatter** during the Big Bang. They want to know more about dark matter and dark energy. And they want to know whether there could be other dimensions to the universe that we do not yet know about.

Did You Know?

Scientist Stephen Hawking will probably lose a bet because of the LHC. He bet a colleague $100 that the Higgs boson would not be found.

Even when the LHC starts up again in 2015, it will take many years to find out all the things that scientists want to know. CERN estimates that so far, the LHC has collected about 1 percent of the data that they need. Scientists think it will take until 2030 to have enough data to know whether their theories about the universe are right. As scientist Dan Green says, "This is a voyage of discovery and we're still in the shallows."

Glossary

antimatter [ANT-eye-matt-err]: Particles that are identical opposites to particles of matter, such that when the two meet, they annihilate each other.

archaeologists [ARK-e-ALL-a-jistz]: Scientists who dig up and study lost civilizations.

Big Bang [big bang]: The explosion that scientists believe occurred at the beginning of the universe.

cathode ray tube (CRT) [CATH-ode ray toob]: A vacuum tube that produces images by projecting a beam of electrons onto a phosphor-coated screen. Old televisions and computer monitors are CRTs.

constituent [kon-STIT-you-ant]: Essential or important parts of a whole.

electricity [e-lek-TRISS-uh-tee]: A form of energy in positive or negative form that can occur in nature or be produced.

electrons [e-LEK-tronz]: Negatively charged particles orbiting the nucleus of an atom.

excavation [ex-ka-VAY-shun]: A digging site.

galaxy [GAL-uk-see]: A group of millions or billions of stars, gas, and dust, held together by gravity.

hadrons [HAD-ronz]: Particles that contain quarks. Protons and neutrons are the most familiar hadrons.

Higgs boson [higgz BOW-sun]: A particle that physicist Peter Higgs believed existed in the Higgs field.

Higgs field [higgz feeld]: A field that physicist Peter Higgs hypothesized particles may have passed through during the Big Bang creating mass.

matter [MATT-er]: Any physical substance that occupies space and possesses mass.

particle accelerator [PART-ick-ull ak-CELL-er-ate-er]: A machine that causes particles to move faster and faster.

physicist [FIZZ-u-sist]: A scientist who studies matter and energy.

superconductors [SOUP-er-kon-DUCK-ters]: Wires that are cooled so electricity can pass freely through them. These electrical currents can make very strong magnetic fields.

three-dimensional [three duh-MEN-shun-ull]: Something that can be measured in three ways: length, width, and depth.

universe [YOU-nuh-verss]: All existing matter and space, everything that is.

voltage [VOLT-edge]: A measurement of electrical energy.

volts [voltzs]: Unit used to measure potential electric energy.

For More Information

Books

Roberta Baxter, *The Particle Model of Matter*. Mankato, MN: Heinemann-Raintree, 2009.

Carolyn Cinami DeCristofano, *Big Bang! The Tongue-Tickling Tale of a Speck that Became Spectacular*. Watertown, MA: Charlesbridge, 2005.

Karen C. Fox, *Older Than the Stars*. Watertown, MA: Charlesbridge Publishing Inc., 2011.

Stephen Hawking and George Lucy, *George and the Big Bang*. New York: Simon & Schuster, 2012.

Jennifer Morgan, *Born with a Bang: The Universe Tells Our Cosmic Story*. Book 1. Nevada City, CA: Dawn, 2002.

Sally Morgan, *From Greek Atoms to Quarks: Discovering Atoms (Chain Reactions)*. Heinemann-Raintree, 2008.

Michael Rubino, *Bang! How We Came to Be*. Amherst, NY: Prometheus Books, 2011.

Alberta Stwertka, *World of Atoms & Quarks*. Harlan, IA: Scientific American Sourcebooks, 1997.

Websites

CERN
http://home.web.cern.ch

LHC Milestones
http://lhc-milestones.web.cern.ch/LHC-Milestones/Flash/LHCMilestones-en.html

Index

About the Contributors

About the Author

Bonnie Juettner Fernandes is a reference book writer and editor living in Milwaukee, Wisconsin.

About the Consultant

Don Lincoln is a physicist at Fermilab and an adjunct professor at the University of Notre Dame. His research uses data taken using both the Fermilab accelerators and the CERN LHC to study the ultimate building blocks of matter and the rules that govern them. Using these scientific marvels, he and his colleagues can study how the universe itself came into existence. He has written many articles and several books on his research, including The Quantum Frontier, a book about the LHC.